YOUR KNOWLEDGE HAS VALUE

- We will publish your bachelor's and
  master's thesis, essays and papers

- Your own eBook and book -
  sold worldwide in all relevant shops

- Earn money with each sale

Upload your text at www.GRIN.com
and publish for free

Bibliographic information published by the German National Library:

The German National Library lists this publication in the National Bibliography; detailed bibliographic data are available on the Internet at http://dnb.dnb.de .

Imprint:

Copyright © 2012 GRIN Verlag, Open Publishing GmbH
Print and binding: Books on Demand GmbH, Norderstedt Germany
ISBN: 978-3-668-11089-2

This book at GRIN:

http://www.grin.com/en/e-book/197810/agribusiness-management-increasing-food-production-with-environmental

Emmanuel Tete Darko

Agribusiness Management. Increasing Food Production With Environmental Consideration

GRIN Publishing

Agribusiness Management:

(Increasing Food Production With Environmental Consideration)

By

EMMANUEL TETE DARKO

July, 2012

Table of Contents   **Page**

Chapter 1: prologue 4-5

Chapter 2: Limitations of Increased And Sustainable Agricultural Food Production.5-7

Chapter 3: Environmental And Social Sustainability in Food Production 7-9

Chapter 4: Forward Looking Approach 9-11

Chapter 5: Conclusion..11-13

Reference..14

List of Figures   **Page**

Figure1: children in Bonsaaso (Ghana) waiting for school feeding......................3

Figure 2: Farmer selling her produce 3

Figure 3: Students in Gumulira (Malawi) receiving school lunch.......................3

Figure 4: Types of food..3

Figure 5: Poor Farm Sanitation..8

Figure 6: Poor Farm Sanitation..8

Figure 7: Unpruned palms...6

Figure 8: Excessive pruning...6

Figure 9: Pocket prunes..6

Figure1: *Kid in Ghana waiting for school feeding* **Figure 2:** *Farmer selling her produce*

Source: *Developed for this work*   **Source:** *Developed for this work*

Figure 3: *Kids in Gumulira receiving school lunch*   **Figure 4:** *Types of food*

Source: *Millennium Promise 2010 Annual Report*   **Source**: *Developed for this work*

Chapter 1: Prologue

"The global population is estimated to grow from the present seven billion to over nine billion in 2050. John Parker inquires whether we will have enough food"
http://www.economist.com/node/18200618

Almost seventy percent of the world's agricultural food production particularly, in crops is produced by a considerable number of smallholders especially in less developed economies. A considering number of these smallholder farmers live in abject poverty, have significantly low productivity, resulting in low incomes and resort to poor methodology in farming that deplete the soil. Forest in the tropic have been destroyed to gratify the need for fertile farm lands, therefore it is imperative that we employ methods that will be sustainable in the long run. Blue chip companies on the globe are conscious of the need to produce sustainably and have shown the commitment towards that. It is in view of this that, Solidaridad and a number of other Non Governmental Organization's as well as multi National Companies decided on the Utz Certified standard for sustainable cocoa in 2009. "At the close of 2009 close to 3,600 smallholders cocoa producers had been trained in certification standards to produce sustainably.
http://www.solidaridadnetwork.org/cocoa

Supplying enough food for the mushrooming global population in environmentally and sustainably tolerable levels, devoid of negotiating on our resources, food production should optimize tremendously from the present production. It should strongly be noted that we can maximize productivity without necessarily damaging the ecology to meet the growing food needs of our time and future. According to the Institute of Food Technologist, (2010 Pg. 574), during the global summit on food security in 2009, it was asserted that, food production must grow between 34%-70% by 2050, if we are to feed the estimated 9 billion people. The obstacle however is the big, increasing food security need in most economies of the world. A large proportion of food produced in less developed and developing economies are not consumed, partly due to poor management and distribution network. In view of this, pronounce malnourishment and dietary deficit in vitamins, minerals, protein, and calories continuous to be common globally. To overturn this dreadful development and ensure a momentous sustainable improvement towards improving the impoverished, this theme should attract the attention of managers of our economies and international bodies

Low cost scientific improvements in food production and distribution mechanisms as well as environmental management and dealing decisively with post harvest management will be key in satisfying this need, indeed failure on our part to hasten the application of science today and yet to be discovered; to food production in an ecologically tolerable level will leave us in critical condition. In fact it will be a fiasco if we increase food production without improving conservation. Variables contributing to food losses differ from one economy to the other as a result of differentials in infrastructure, investments and awareness levels

The milestone of this paper will concentrate on agriculture primary food production since food production encompasses a wide range of produce, thus all issues will be constricted to improving farm produce to meet the growing demand of the global market, whiles considering the sustainability of such produce with the view to ensuring exponential increase in farmers' incomes.

There are four over-arching goals in writing this paper. The first, and most basic, is discussing challenges confronting increased sustainable agriculture food production The second is discussing the underpinning sustainability by restoring and conserving the natural resource base, managing the ecological and societal sustainability issues connected to the development of food crop production by optimizing performance in existing farms and improving the income levels of smallholder producers.

The third objective is to discuss hopeful but innovative channels of forward looking approach by presenting a promising way of increasing food production sustainably.

The spotlight of this essay will begin with a prologue that abridge the growth of cultivation, food knowledge and sustainability perspective; a section on challenges confronting optimum crop production to meet the needs of humanity; a section about potential environmental and social consideration solutions; a section on forward looking approach and concluding remarks.

Chapter 2: Limitations of Increased And Sustainable Agricultural Food Production

"World population is expected to rise by 36% between 2000 and 2030. Average crop yields are anticipated to grow at about the same rate. However, relative yield increases have been undermined generally (projections for global yields in the next decade predict 1-1.1% p.a. growth for cereals)."
http://www.sefalliance.org/fileadmin/media/sefalliance/docs/Call_Seminar_Downloads/2 0100311_Assessing_Biofuels_Presentation.pdf

The reasons are farfetched; weak agronomic practices in the form of poor: production, maintenance of farm, and post harvest management as well as international charters have been the bane for most smallholder producers in sub-Saharan Africa.

With respect to production, most often than not, the soil is not prepared and where it is done, the preparation is woefully executed, "Africa is the lone constituency on the globe where per capita food productivity has been waning for the past three decades. Cereal yields in Africa are a quarter of the international average and Africa's soils remain the worst globally." Karapinar, Baris Haberli Christian (2010, Pg. 82); most smallholder farmers adopt low yielding planting materials, and where they are used, they are not appropriate for the soils and climate; more imperatively, depleted soil fertility; absence of small irrigation systems due to over reliance on rain, Crop plants require a continuous supply of water to replace the water evaporated (transpired) from their aerial organs; irresponsible use of agro-chemicals; non observance of proper planting spacing that are necessary within the interrows immensely impinge on harvesting paths, weed control, fertilization and other agronomic practices; the high planting density in planting pattern leads to competition for basic crop requirements. On the other hand, a sparse planting density (evident in some farms) is a waste of land resources. For good growth and productivity, the optimum spacing for every food crop farm should be observed to circumvent crop plant from being tall and etiolated.The constant attack of insect, pests and diseases on farms all contract yields; ill timed harvesting periods are equally contributing variable to low productivity, thus long and short harvesting cycle if not appropriate to the food crop affect the quality of the produce to a larger extent. International bodies such as the World Bank, IMF and other powerful bodies have via policies contributed to low productivity unconsciously, the inexorable of such policies and financial support to developing economies is migration of needed farm labor in food producing areas to the urban slums. Regarding farm maintenance, "the maintenance regime (weeding, pruning) on most farms are generally poor. There were cases of excessive pruning which tended to affect the architecture of the palms. An instance of fronds supporting developing bunches being cut was observed. None of the farms had been planted with cover crops and there were no harvesting paths" CSIR-OPRI, (2012). The figures below are illustration of poorly maintained oil palm farm at Bonsaaso Ghana

Figure 5: *Poor Farm Sanitation*   **Figure 6:** *Poor Farm Sanitation*

Figure 7: *Unpruned palms*   **Figure 8:** *Excessive pruning*   **Figure 9:** *Pocket prunes*

Source: Developed for this work

Concerning post harvest management, smallholder farmers in the sub-Saharan African do not have storage facilities and with market vulnerabilities confronting them most farm produce are left to go waste. Thus non guaranteed prices for smallholder producers will inexorable contract production levels. Poor infrastructural network compound this menace and further waning their prospects to make money. At the root of the problem is the high cost of fertilizer and limited access to them.

Chris Pollock, Jules Pretty, Ian Crute, Chris Leaver, and Howard Dalton in a publication in the Royal Society contended that estimation of climate change impacts propose that there will be significant adverse effects on crop production, predominantly in sub-Saharan Africa. http://rstb.royalsocietypublishing.org/content/363/1491/445.short

Lawrence Geoffrey, Lyons, Kristen, Wallington, Tabatha. (2010, Pg. 3) posit that droughts, floods, disease, pestilence and other so called natural disasters have ceaselessly affected the level of food accessible for human use. Such events will persist to impact negatively on agriculture

Chapter 3: Environmental And Social Sustainability in Food Production

The popular axiom states that 'the last man dies when the last tree dies' and according to Perfecto, Ivette, Vandermeer, John H., Wright, Angus Lindsay, (2009, Pg. 11), the world will surely be diminished as the last wild gorilla is shot by a local warlord grateful to one or another political ideology, or as a rare but gorgeous bird species has its habitat detached to make way for yet another desperately needed line of supermarket.

societal, green (ecological) and commercial factors are entrapped, generating volatility; ecological anxiety such as depleting of reserves, coupled with communal matters akin to the volatile increase in population in developing economies, are joining forces in a sparkling cauldron. For sustainability to be attained, then marketplace should be the centre piece, put differently, sustainability requires fully the assistance of market forces, i.e. firms and manufacturers. And it must be toughened by development at the marketplace; this encapsulates pricing that will be reflective to the social-ecological costs of production and green standards. Corporate social responsibilities, good governance on the part of governments and deep participation of civil society are the three pillars of developing sustainable market. Sustainable food production can make a noteworthy input to fighting deficiency and conserving the natural setting. Today sustainability is on the program of numerous firms that accept accountability for the source of their goods and are now witnessing sustainability as a requirement for the permanence of their trade. Environmental and social sustainability improve marketplace penetration with a competitive price for a quality product, thus providing durable interaction with purchasers and enjoying gratitude and incentive for sustainably produced goods.

Increasing environmental sustainability in food production reinforces the sovereign place of smallholder producers and explains the need for capacity building as professional's of farming practice and operational running. It is envision that by this their produce will be enhanced qualitatively and offer them optimum productivity at low operational expense which invariably positions them strongly in negotiating for realistic price to reflect the ecological cost of production. This is a sure way to improve standard of living.

To gratify the exponential increase in global preference for food and the call for sustainable production, food producer groups need to be supported in building the evolution to people and environmentally welcoming farming. Interventions that offer smallholder producers an exit of the discouraging coiled of poor farming techniques, worn out soils, low harvest and low incomes. Building farmers capacity in sustainable farming techniques results in significant excellence produce and boost incomes. This modus operandi considerably lessens unpleasant impact on the environment. This christen for collaboration from all stake holders along the food production chain.
As part of this, smallholder producers should be orientated to desist from the use of child labor on their farms and all farm support to farmers' organization should be tied to those organizations that observe this.

Innumerable agriculture techniques are not only untenable, but suboptimal. Most farm space under cultivation are operating at excess capacity that is, the areas under production can yield three to five times more per hectare if scientific applications are employed. It is imperative to assert that in the decade or so food production for alarming global population will need to increase significantly – and clearance of forests with the view to improving productivity will not be an option. Thus delivering food security globally and particularly to the vulnerable calls for the need to use land prudently: prioritizing productivity, superior quality and production at extremely reduce costs per unit of production. Doing extra with less – optimum production with far less impact on biodiversity, reduces usage of water and energy, less emissions, and reduce fertilizer and agro chemical inputs. This evolution to sustainable agriculture, superior concentration should be given to effective farm administration and agricultural entrepreneurship, whiles Fostering close knot with determined ecological and societal program.

Resource protection and regenerative know-how are blatantly diffusing. "In Denmark, some 150 farms have in-field weather stations to facilitate disease outbreaks prediction resulting in cuts in agro-chemical application; in UK, about 150,000 hectares of farms were computer-mapped in 2000s, allowing inputs to be targeted more accurately and the sum application of agro-chemicals to be slash and also in the same UK three-quarters of crops grown in glasshouses use natural predators to control pests rather than pesticides". Pretty, Jules N. (2008, Pg. xxxv)

The issue to be concerned with is, will improvement in sustainability result in higher food productivity? Adopting integrated farming practices could immensely promote sustainability targets and reduce farm expense on inputs. Integrated farming practices incorporates organic farming, pest management, and health concerns and has extremely positive effects in the sense that produce from such practices are far better and command premium price due to the less application of agro-chemicals. It also has available technology to keep natural resources for a long period of time; it is insect resistance and balance between natural and artificial process, as well as stabilizing the soil nutrient content. Notwithstanding this, integrated farming practices demand more knowledge and seem to be expensive in the short run

Chapter 4: Forward Looking Approach

The benefits inherited in emerging out of predicament and issues in agric marketplace are decisive for the global poor. There are valuable dividends in facilitating brisk movements towards sustained growth in famine reduction. With the viewpoint of global market remaining doubtful, farming markets reservation has heighten over time, making it vague. In brief considerable ambiguities continue in agric marketplace across the globe. Away from the superseding query of the time as well as the pace of revival from the cruel economic downturn, some concerns especially to agriculture and its marketplace seem grave for potentially international market.

Against the backdrop of facilitating optimum productivity, a number of producer schemes need to be introduced, including input subsidies, access to financial institutions and credit, output price support, linkages to global markets, research and extension services made available to farmers by incorporating farmers inputs, small scale water management initiatives to drip irrigation for high value crops, crop insurance programs, training farmers in effective agronomic practices and certification, provision of community storage facilities, high yielding planting material, adherence to proper farm sanitation practices, proper application of agro-chemical.

Programs for sustaining partial subsidies for basic agricultural inputs are essential in optimizing production levels to meet the global food needs. It is imperative to assert that input subsidies may only raise input prices as supply could be less than demand for it and this will profit suppliers of such inputs instead of smallholder producers since it could raise farm cost of production and eventually affecting output prices. Subsidy for seed, equipment and energy for machinery and equipments including fertilizers and high value yielding planting materials to farmers will be crucial intervention for high production levels which invariably will increase farmers' income, a necessary ingredient in combating poverty.

Strengthening farmers' capacity for technology development is central to farmers' first strategies. There is growing evidence that farmer-first approaches can succeed. Key to this success are farmers' organization's capacity to diagnose and prioritize their own needs; to access a 'basket' of technological options; and to test, evaluate and adapt technologies. Training smallholder farmers to consider farming as a serious business, the establishment of farmer business schools to train farmers in improved management techniques (timely planting, plant spacing, seed and fertilizer placement, shade management, pruning, disease management, harvesting and post harvest management and construction of low cost storage facilities that have good aeration and keep out rodents and other pest. Improved storage will help farmers to organize and negotiate for competitive prices for bulk sales of their produce. This training should be a concerted effort of agric extension department; Ngo's and credit institutions and research bodies.

Strategies aimed at improving finances and credit accessibility are essential for smallholder farmers to make the needed investment on their farms and even to diversify to support the transition from subsistence farming to commercial agriculture. To be able to access this credit without much difficulty, credit institutions should persuade farmer groups to embrace micro crop insurance programs either as part of the credit scheme or total farm insurance that is a sure way of circumventing potential fear of repayment in future. Formation of farmers cooperatives should be facilitated by Ngo's that operate in farming communities and whose core business it is to battle poverty, such groups and cooperatives will significantly facilitate easy access to credit from the financial institutions and could easily link their produce to global market with the view to gaining competitive price that will reflect the social and environmental cost of their produce. Thus, higher farm income, improved standard of living greater investment on farms and the cycle continues.

Imperatively, once smallholder farmers are organized, they could be better targeted by these credit institutions, the later can integrate their operations with farmer groups and big buyers of farmers produce and that will allay any repayment default possibilities because the produce buyers will be making payment of produce bought into the account of farmers with the credit institutions and this will assure the credit institution to be courageous to lend to these farmer groups. This however needs to be crafted and implemented cautiously to make it sustainable.

As Africans depend on rain-fed agriculture for staple crop production for the foreseeable future, yields are subject to inter- annual rainfall variation together with the negative upshot of atmospheric transformation. One crop failure can wipe away the gains made in several years. Interventions that build resilience and contract potential risk should be implemented- from small scale water management initiatives, to drip irrigation systems for high value crop. Once farmers group themselves, other stakeholders could facilitate the construction of this technology with the view to Lessing the dependency on rain fed farming.

"Mr Paul Ainoo, Chairman of the Cape Coast Farmers Cooperative Society limited has appealed to the State to craft a policy framework to set up an organization responsible for the advancement of citrus and lime sector with the view to preventing the sector from entire disintegration. Asserting that inadequate robust market and an institution to run the sector, has led many farmers to either fell their citrus trees or has in total abandoned their farms for different source of revenue; even as some were thinking of discarding citrus farming for other ventures. He opined that citrus farmers in spite of inadequate buyers for their produce still engage labour to work on their farms. In view of this, he calls on the government and all partners in the business for the formation of "A citrus Board" to restructure all citrus farming actions, stating that a board will enhance upon the negotiation influence of farmers."
http://news.peacefmonline.com/social/201111/81818.php

From Paul Ainoo's lamentation above, it is crystal clear that guarantee market with output support price will be a sure way of ensuring sustainable food production since farmers need to sell all their produce with the view to paying back any credit whatsoever contracted to finance their production if any. Apprehension over dependability of global market as a basis of food provisions has refurbished the focal point in multi economies on food self satisfactoriness as a conduit of attaining food protection. Couple of net food importers around the globe are fine-tuning their strategic agricultural policies and rather prioritizing optimum food production to reverse this unfortunate trend

For instance "The Philippines are resolute to encourage food production with the view of attaining self-sufficiency in staple foods by 2010. Armenia announced an attempt to reach self-reliance via subsidies for expansion of cropland and irrigation. The Government of Kazakhstan planned to introduce US$3 million into the agriculture sector to support farmers endure the shock of the global financial turmoil. Malaysia allocated US$1.29 billion to increase rice-growing while also maximizing government minimum prices for rice". www.fao.org/catalog/inter-e.htm

Transforming peasant food production however requires a fundamentally different approach to research and extension. Both government agencies and non-government organizations must enable smallholder farmers to become the principal agents for change. Agricultural research and extension institutions need to be more responsive to farmers' knowledge and innovation and allow farmers to be creative analysts and experimenters rather than passive recipients. Contemporary science can contribute to this process as a partner and resource, rather than as a source of blueprint for new technologies. Farmer organizations can provide crucial help in shaping more effective research and extension for complex, diverse and risk-prone areas. Smallholder farmers' cooperatives can link research and extension agencies with farmers' knowledge, innovative capacity and expectations. The cooperatives should adapt and disseminate agricultural technologies in programs they themselves manage and control.

Feeding the population globally though require the expansion of global cropland, expansion of farm size with the view to accelerating production efficiency only results in deforestation and endangers biodiversity; rather science based methodology should be employed to obtain the optimum yield from the same farm size.

Daniel Imhoff, Fred Kirschenmann, Micheal Pollan, (2012, Pg. 10) contended that we need food and farming systems that share our limited planetary resources so that citizens in every region of the planet can become food self-reliant. Award programs and certification systems will provide a powerful tool, incentives and framework for producing food crops in an environmentally tolerable manner. Certification schemes such as Utz Fair-trade etc should be applauded and strengthened.

Conclusion

Impoverished persons on the globe as a result of growing need for boom food production could only have some hope if there is extensive application of science and technology in our farming methods and this will invariable reduce the food gap. The vast majority of smallholder farmers in sub-Saharan Africa remain unorganized. Without strong, community based organization, it will be extremely difficult if not improbable to see how smallholder groups can have enough clout to act as a credible and effective partner with stakeholders. The essence of smallholder farmers coming together in the form of cooperatives and other groups is to facilitate the application of technology to obtain pareto optimality from their farming activities. This should be manifest in mapping of their farms to determine the actual farm size which will be key in the quantum of agro-chemicals to apply on the farm, the number of planting materials to be planted on the same piece of farm land, to enhance the facilitation of credit facilities from the financial and other credit institutions and also to smooth the progress of yield estimates.

Weak agronomic practices in the form of poor: production, maintenance of farm, and post harvest management have been the bane for most smallholder producers in sub-Saharan Africa; depleted soil fertility; absence of small irrigation systems due to over reliance of rain; irresponsible use of agro-chemicals; non observance of proper planting spacing that are necessary within the interrows immensely impinge on harvesting paths, weed control; fertilization and other agronomic practices; the high planting density leading to competition for basic crop requirements; the maintenance regime (weeding, pruning) on most farms are generally poor; constant attack of insect, pests and diseases on farms contract yields; ill timed harvesting periods (long and short harvesting cycle) are equally contributing variable to low productivity; non guaranteed prices for smallholder producers which contract production levels; poor infrastructural network among others have contributed in diverse ways in providing food insecurity to the rapidly increasing global population today.

The decline of public extension services is one of the most striking changes in the agricultural landscape over the past decade since farming systems worldwide have been going through dramatic changes as a result of globalization, liberalization and rapid urbanization. Yet without functional research-extension architecture in place, the hoped-for improvements in agronomic practices by tens of millions of smallholder farmers are unlikely to materialize; reversing decline in soil fertility with subsidizes mineral and organic fertilizers, providing improved crop varieties appropriate for soil and climatic conditions, establishing low cost storage technology, financing merge of donation, credit and guarantees for self-sufficient, will ensure quick wins in optimizing food production to feed burgeoning global population.

The sustainability discussion is mounting in strength as scientists churn out incompatible knowledge as well as resistance from interested locations and philosophies become focal point. The task is to display that sustainability brings variations – a change to producers' incomes, to operational environment, to the ecology, and accessibility of uncooked resources and agricultural harvest for the next global stage.

 Furthermore, most initiatives that are under deliberation and aiming at sustainable production intensification' involve complicated combination of domesticated plant variety and allied organization method, demanding superior skills and awareness by smallholder producers and the agribusiness community. To achieve high production efficiencies and sustainably, smallholder producers need to recognize the circumstances under which agricultural inputs can either harmonize or oppose biological progression and flora and fauna services that intrinsically hold up agribusiness systems.

Signals of burst through are not rare. The moment to systematize the switch towards a sustainable economy has emerged, innovative preparations need to be fashioned, and coalitions need to be forged and functioning processes rationalized. A vital thought in this is scaling up: what has previously been experienced and established to be effective need to be up scaled and made extensive applicable and reachable. This demanding mission is in a decisive chapter. Can we ensure swift sufficient advancement with generating support for novelty, injecting fiscal capital in the correct location and constructing the tenacity needed to demonstrate a different approach?

Execution of sustainability principles has confirmed to be a constructive instrument for leveraging effective agricultural and social performances and improving productivity; Sustainable food production decreases scarcity and can diminish heaviness on jungle with lofty social, cultural and natural morals.
The potential of farming and food production principally will be firmed up by the line of three chief inclinations: people and related demography; accessibility and kind of power possessions; environmental manipulation of obtainable land, water, and atmospheric excellence. Imperatively, population being the core issue and coerces others (given that it exerts force on need for land) via its multiplier, the level of comfort. On the other hand, the technologies arranged will also be a subject of technology consideration, open strategy, buyer behavior, and financial wealth.

It is insightful to opine that, the forward looking approach should be combined in an integrative comportment to create synergistic results, if we are to attain the gains associated with their employment on arable land. When employed in dis-jointed effort, could as well obtain the same unfavorable outcomes when ignored completely.

AS , Ingram, Helen M., Garrido, Alberto (2011, Pg. 14), put it "there seems to be significant uncertainty in meeting future world food demand for the next two decades, and there is even more uncertainty in achieving the goal of eradicating or reducing the hunger that presently affects more than 800 million people worldwide" Yes, natural phenomenon won't be less burdensome on productivity but there is the need for us to alter our lens from natural disasters in order to appreciate the structural variable underlining high food production. The multimillion question is, will industrial growth happen without considerably growth in foodstuff productivity?. Weaknesses demonstrated in farming activities in Sub-Saharan is the basis of the sluggish fiscal development

Reference.

CSIR-OPRI, (2012) Report On Inspection Of Oil Palm Farms And Selection Of Suitable Nursery Sites For Sustainable Oil Palm Production. CSIR-OPRI/TR/S.O.A/2012/85

Daniel Imhoff, Fred Kirschenmann , Micheal Pollan (2012) Food Fight: The Citizen's Guide to the Next Food and Farm Bill. Watershed Media

Godfray HCJ, Beddington JR, Crute IR, Haddad L, Lawrence D, Muir JF,Hitchcock, Darcy E., Willard, Marsha L. (2009). Business Guide to Sustainability: Practical Strategies and Tools for Organization 2nd Edition. Earthscan Ltd.

Ingram, Helen M., Garrido, Alberto (2011). Water For Food in a Changing World. Routledge

Karapinar, Baris Haberli Christian (2010). Food Crises And The WTO: World Trade Forum. Cambridge University Press.

Lawrence Geoffrey, Lyons, Kristen, Wallington, Tabatha. (2010). Food Security, Nutrition And Sustainability. Earthscan.

Perfecto, Ivette, Vandermeer, John H., Wright, Angus Lindsay, (2009). Nature's Matrix: Linking Agriculture, Conservation and Food Sovereignty. Earthscan.

Institute of Food Technologists, (© 2010). Comprehensive Reviews in Food Science and Food Safety _Vol. 9,

Pretty J, Robinson S, Thomas SM, Toulmin C. (2010). Food security: the
Pretty, Jules N. (2008).Sustainable Agriculture And Food. Vol.1. Earthscan Ltd.

http://news.peacefmonline.com/social/201111/81818.php

http://rstb.royalsocietypublishing.org/content/363/1491/445.short

http://www.economist.com/node/18200618

http://www.sefalliance.org/fileadmin/media/sefalliance/docs/Call_Seminar_Downloads/2
0100311_Assessing_Biofuels_Presentation.pdf

http://www.solidaridadnetwork.org/cocoa

www.fao.org/catalog/inter-e.htm

YOUR KNOWLEDGE HAS VALUE

- We will publish your bachelor's and
  master's thesis, essays and papers

- Your own eBook and book -
  sold worldwide in all relevant shops

- Earn money with each sale

Upload your text at www.GRIN.com
and publish for free